Seed Testing Techniques for Seed Spices

About the Authors

Dr. (Mrs) Sangeeta Yadav, born on April 27, 1980 was awarded B.Sc. in Agriculture and M.Sc. (Ag) Horticulture degree in 2005 and 2007, respectively from Jawaharlal Nehru Krishi Vishwa Vidyalaya, Jabalpur and secured first division in both the examinations. She was awarded Ph D degree from Barkatullah University, Bhopal in 2014. She was actively involved in co-curricular activities during her education and bagged several awards in sports, "B" certificate in NSS and got state level prize on Mango day by Chief Minister of Madhya Pradesh.

She has earned specialization on seed spices in her Master and Ph D research work. She has to her credit eleven scientific research papers in referred and fourteen in non-referred Indian and foreign journals. She is an author of a bulletin on 'Coriander: Status and Prospects'. She is serving as the referee of 'Turkish Journal of Botany'. She is member of nearly half dozen professional societies and providing consultancy to seed spice growers. She is promising, young and growing scientist who is extremely devoted for the development of seed spices in the country and livelihood security of farming community.

Dr. Arun Kumar Barholia, born on January 01, 1956 is presently serving as Senior Scientists (Horticulture) in department of Horticulture, Rajmata Vijayaraje Scindia Krishi Vishwa Vidyalaya, College of Agriculture, Gwalior, M.P. He was meritorious student since beginning and completed his graduation in agriculture and post-graduation in horticulture from Jawaharlal Nehru Krishi Vishwa Vidyalaya, Jabalpur. He was awarded Ph D degree from Barkatullah University, Bhopal in 2005.

He has long experience (36 years) in teaching, research and extension activities in horticulture. He has to his of 22 research papers in referred journals, 51 in non-referred journals and magazines. He is an author of three published booklets on grape, banana and chilli. He has supervised 08 P.G. and 01 Ph D. student. He is also credited for delivering more than 35 radio talks and 45 television talks. He is devoted for upliftment of farmers who are engaged in growing of horticultural crops.

Seed Testing Techniques for Seed Spices

Sangeeta Yadav
B.Sc. in Agriculture and M.Sc. (Ag) Horticulture
Jawaharlal Nehru Krishi Vishwa Vidyalaya
Jabalpur, M.P.

Arun Kumar Barholia
Senior Scientists (Horticulture)
Department of Horticulture, College of Agriculture
Rajmata Vijayaraje Scindia Krishi Vishwa Vidyalaya
Gwalior, M.P.

NEW INDIA PUBLISHING AGENCY
Pitam Pura, New Delhi – 110 034

New India Publishing Agency
101, Vikas Surya Plaza, CU Block, LSC Market
Pitam Pura, New Delhi 110 034, India
Phone: + 91 (11)27 34 17 17 Fax: + 91(11) 27 34 16 16
Email: info@nipabooks.com
Web: www.nipabooks.com

Feedback at feedbacks@nipabooks.com

ISBN : 978-93-85516-91-7

Composed, Designed and Printed in India

राजमाता विजयाराजे सिंधिया कृषि विश्वविद्यालय
रेस कोर्स रोड, ग्वालियर (म.प्र.) – 474002
Rajmata Vijayaraje Scindia Krishi Vishwa Vidyalaya,
Race Course Road, Gwalior (M.P.) – 474 002
(An ISO Certified 9001:2008)
Tel: 0751 – 2467673, Fax: 0751 – 2464141, E-mail: vcrvsaugwa@mp.gov.in

प्रो. अनिल कुमार सिंह
कुलपति
Prof. A.K. Singh
Vice-Chancellor

No./VC/2016-17/2359
Date: 22/10/2016

Foreword

The most remunerative commodities of the arid and semi-arid regions of India are the seed spices, which are a high value low volume crops. This group of spices comprise of major crops like coriander, cumin, fennel, fenugreek and ajwain. Now nigella, dill, celery, anise, caraway etc. are emerging as minor seed spices. India being largest producer, consumer and exporter of seed spices, can exploit this opportunity through development of these crops and improve the livelihood of their growers.

The present publication entitled, "Seed Testing Techniques for Seed Spices" is targeted to meet the long awaited demands of researchers, seed technologists/ technicians, seed laboratories, students, growers and all other stakeholders who are engaged in overall developments of seed spices in the country.

The sincere efforts made by authors of this publication Dr. Sangeeta Yadav and Dr. Arun Kumar Barholia are commendable and praiseworthy.

A. K. Singh

Preface

India is known as the 'Home of Spices' in the world over. Potential for export and trade of seed spices are growing globally. India avails the opportunity of being largest producer, consumer and exporter of seed spices. Spices constitute an important group of agricultural commodities, which have been considered indispensible in the culinary art for flavoring foods. Some are also used in pharmaceutical, perfumery, cosmetics and several other industries. In India, about 52 and 31% of total seed spice production comes from the states of Rajasthan and Gujarat, respectively. Among the seed spices, coriander, cumin, fenugreek and fennel are the major seed spices, whereas, ajawain, dill and few other have been categories as the minor seed spices.

Looking to quite a significant role in our national economy, it is imperative to assess the seed quality with respect to field and laboratory standards as per Seed Act and ISTA rules. It is considered to be a boon in increasing the quality production and productivity of seed spices in the country. Keeping this view point in mind, an endeavor has been attempted to bring all these necessary information in the form of book entitled "Seed Testing Techniques for Seed Spices" which will fulfill the long awaited need of researchers, students and progressive farmers.

Authors express sincere thanks of gratitude to Prof. Anil Kumar Singh, Vice Chancellor, RVSKVV, Gwalior and Prof. Vijay Singh Tomar, Vice Chancellor, JNKVV, Jabalpur for their encouragement and supports. We also wish to express thanks to Dr. S.S. Tomar, Dean, Faculty of Agriculture, Dr. H.S. Yadava, Director Research Services and Dr. S.K. Srivastava, Director Extension Services, RVSKVV, Gwalior for their cooperation and help during the preparation of the manuscript. Thanks are also due to those scientists whose published works have been referred.

Gwalior, Jan. 2017

Dr. Sangeeta Yadav
Dr. Arun Kumar Barholia

Contents

1

Introduction

India is famous as the '*Land of Spices*' world over. Spices are nature's gift to humankind for flavoring our food. They are also used as preservative and in pharmaceuticals. India is the largest producer, consumer and exporter of seed spices and their value added products. About 63 different spice crops are cultivated in the country. Among them, seed spices are annual herbs, whose dried seeds or fruits are used as spices. These crops share 21.87% in terms of quantity and 7.42 % in terms of value to total spices export in the country. The country grows a wide variety of seed spices like coriander, cumin, fenugreek and fennel as major crop, while, dill, nigella, ajwain, anise, celery, caraway, parsley, black cumin and few other are categories as the minor seed spices.

In India, nearly 12.1 lakh tones of seed spices namely, coriander, cumin, fenugreek, fennel, ajwain, dill and celery are produced from an area of about 13.8 lakh hectares. The states of Rajasthan and Gujarat are major seed spices growing Indian states and contributes about 52 and 31% of total seed spice production of the country, respectively. Madhya Pradesh ranks second in production of coriander with 17% contribution in national coriander basket. Rest of the seed spices are also grown in considerable area and contribute sufficiently in national production of these crops. However, the realized productivity of these crops is below the expectation and extension gaps exist. It is therefore urgently felt to develop high yielding varieties with better quality and improved technologies for increasing seed spices production to fulfill the demand of rising world population.

In India, most of the cultivable land (< 50.0%) is facing the problems like soil and water degradations, water shortage for cultivation, soils salinity, alkalinity, macro and micronutrient deficiency, heavy metals and various pollutants contamination. Seed spices have potential to tolerate above problem to certain extent. The saline/alkaline soils may be utilized for the production of fennel and dill. All the seed spices are tolerable to mild soil acidity and alkalinity whereas, coriander and fenugreek can come up well even up to 5.5 soil pH levels. Coriander, cumin and fenugreek are suitable for nutritionally eroded soil. These crops can be cultivated in rainfed farming in moisture deficit conditions.

Quality seed is the most vital and critical input in crop production. Use of quality seeds of improved varieties, suitable to location specific condition has played significant role in enhancing crop production and productivity in the country. It is, therefore, imperative to place more emphasis for developing efficient, effective and latest technologies of seed production, which should relatively be of low cost and affordable to low-income farmers. The replacement of existing varieties by new ones is necessary for better yield and profitability. The percentage of area sown by using certified seed to total cropped area in a season is called seed replacement rate. Ideally, seed should be replaced every year for hybrids and every three to four years for improved varieties of self-pollinated crops. However, in practice, seed is replaced less often in case of open pollinated crops. As a consequence, seed replacement rate is lower than recommended for different seed spices crops. In field conditions in general, the recommended package of practices do followed as such and variation results in the deterioration of improved varieties. The major reasons for deterioration of genetic purities of crop varieties are mentioned below:

- Developmental variation
- Mutations
- Natural crossing
- Latent genetic variability
- Selective influence of disease and pest
- Mechanical mixture
- Techniques of plant breeders

It is therefore becomes important to maintain the genetic purity of recommended varieties of seed spices by a sound maintenance breeding programme and produced the demanded quantity of certified seed to meet the farmers demand. The strategies for quality seed production in seed spices should be based on the followings:

- Identification of current cropping pattern
- Current seed scenario
- Fixing the targets for SRR
- Seed demand assessment

- Selection of seed chain varieties
- Preparation of seed rolling plan
- Checking the points of seed deterioration
- Effective monitoring during seed production
- Strengthening of quality control system
- HRD in seed production
- Increasing seed awareness

Similarly, a sound seed rolling plan can be prepared for the production of certified seed in order to maintain the demand and supply chain in seed spices. The point's need attention in the preparation of seed rolling plan is listed below:

- Cultivated area
- Requirement of certified seed
- Farming conditions (Rainfed/Irrigated/Late Sown-Kharif/ Rabi/ Spring/ Summer)
- Selection of crop varieties for each farming situation
- Estimation of share of varieties
- Current seed replacement rate (SRR)
- Targeted seed replacement rate (SRR)
- Seed multiplication ratio (SMR)
- Seed rate per hectare
- Estimation of requirement of different kind of seed like nucleus seed, breeder seed, foundation seed and certified seed.

The quality of seed in accordance to Indian Seed Standards is judged with the help of physical purity test, germination test, tetrazolium test, viability test and seed moisture tests, however, the information on techniques used for these tests are not much familiar for the seed spices which are multiplied mainly by the seeds. The various techniques to be used for seed testing in the case of seed spices to assure the physical purity are described in different chapters.

2

Notified Varieties

Notified Varieties of Different Seed Spices

1. Coriander

S. No.	Varieties	Notification status	Pedigree	Maturity days	Average yield (kg/ha)	Reco-mmended ecology	Adapta-bility states
1	Aphu – Dhana (LCC 170)	1979 (E) 20.08.2010	Selection from land race of Kurnool district	75 - 95 days	1000 -1200 kg/ha	Rainfed conditions	Tamil Nadu, Haryana and Bihar
2	Hisar Sugandh (DH 36)	597 (E) 20.04.2006	Mass selection from germ - plasm	130 - 140 days	1400 - 1500 kg/ha	Limited irrigation conditions	Central zone of the coun-try
3	Pratap Raj Dhania 1(RKD 18)	268 (E) 28.01.2015	-	100 - 110 days	1700 - 1900 kg/ha	Irrigated conditions	Rajasthan
4	Punjab Sugand	268 (E) 28.01.2015	-	120 - 130 days	1500 - 1800 kg/ha	Irrigated conditions	Punjab
5	RCR 435	597 (E) 20.04.2006	Recurrent selection	115-125 days	1000 -1250 kg/ha	Irrigated high fertili-ty condi-tions	Rajasthan
6	RCR 436	664 (E) 11.05.2005	Recurrent selection	120-125 days	1000 - 1250 kg/ha	Rainfed conditions	Rajasthan
7	RCR 728 (UD 728)	1979 (E) 20.08.2010	Recurrent selection	130-140 days	1370 - 1500 kg/ha	Rainfed conditions	U.P., Rajasthan & Bihar
8	Sudha	858 (E) 01.06.2007	Selection from germ-plasm	56 - 80 days	1400 - 1700 kg/ha	Rainfed conditions	Rajasthan

2. Cumin

S. No.	Varieties	Notification status	Pedigree	Maturity days	Average yield (kg/ha)	Recommended ecology	Adapt-ability states
1	Gujarat Cumin 4 (GC 4)	2035 (E) 28.11.2006	Selection from GC 3	100 - 150 days	700 - 800 kg/ha	Rainfed conditions. Resistant to wilt & Powdery mildew	Gujarat
2	RZ 223 (UC 223)	858 (E) 01.06.2007	Mutant of UC 216	120 - 130 days	200 - 300 kg/ha	Rainfed conditions.	Rajasthan & Gujarat

3. Ajawain

S. No	Varieties	Notification status	Pedigree	Maturity days	Average yield (kg/ha)	Recommended ecology	Adapt-ability states
1	Azad Ajawain 1	2035 (E) 28.11.2006	-	160 - 170 days	1100 - 1200 kg/ha	Rainfed farming Resistant to root & stem rot	U.P.
2	Pant Ruchika	2035 (E) 28.11.2006	Pure line selection	150-180 days	700 - 1600 kg/ha	Rainfed as well as irrigated conditions	U.P. & Uttaranchal

4. Fennel

S. No.	Varieties	Notification status	Pedigree	Maturity days	Average yield (kg/ha)	Recommended ecology	Adaptability states
1	Gujarat Fennel 11 (GF 11)	2035 (E) 28.11.2006	Single plant selection from germplasm	150 – 160 days	1500 – 1700 kg/ha	Irrigated conditions	Gujarat
2	RF 101	664 (E) 10.05.2005	Recurrent selection (Half sib)	150 – 160 days	1400 – 1700 kg/ha	Suitable for loamy & black soils. Resistant to Rumularia blight & Alternaria blight	Rajasthan
3	RF 125	597 (E) 20.04.2006	Recurrent selection (Half sib) in EC 243380	150 – 160 days	1500 – 1700 kg/ha	Suitable for irrigated conditions of loamy & black soils.	Rajasthan
4	RF 143	858 (E) 01.06.2007	Recurrent selection from local germplasm	145 – 155 days	600 – 700 kg/ha	Rainfed conditions	Rajasthan & Gujarat
5	RF 205	1979 (E) 20.08.2010	JF 25 x RF 125	140 – 150 days	1600 – 1800 kg/ha	Tolerant to Rumularia blight	Rajasthan, U.P. & Bihar

3

Seed and Seed Chain

Seed

The seed, in botany is a mature ovule, consisting of an embryo together with a store of food all surrounded by a protective coat and bear ability to develop a new plant on optimum conditions of soil, moisture and light. It usually develops after the fertilization on an egg cell by male generative cell from a pollen grain. Seeds of some species develop without the intervention of the male cell, formed entirely of 'mother' tissue. Such seeds are called apogamic seeds. In seed technology, the seed is seed or any vegetatively propagated material produced from seed crop in accordance to Indian seed standards and has potential to develop new crop in presence of soil, moisture and light.

Classes of Seed

i. Nucleus seed

ii. Breeder seed

iii. Foundation seed

iv. Certified seed

i. Nucleus seed

This is the cent – per cent genetically pure seed with physical purity and produced by original or sponsor breeder / institute or SAUs from a basic nucleus seed stock. A pedigree certificate is issued by the producing breeder.

ii. Breeder seed

Breeder seed is the progeny of nucleus seed produced under the supervision of a plant breeder/research institute / Agricultural University and its production is monitored by a joint inspection team of scientists and officials of certification agency and Seed Corporation. This is also 100 per cent genetically pure seed as to guarantee that in the subsequent generation of foundation and certified seed classes shall conform to the prescribed standards of genetic purity. The other quality factors such as physical purity, inert matter, germination etc. are indicated on the golden yellow colour label on actual basis for this category of seed.

iii. Foundation seed

Foundation seed is the progeny of breeder seed which is produced by recognized seed producing agencies in both public and private sectors under supervision of qualified technical officers of certification agency in such a way that its quality is maintained according to prescribed seed standards. A white colour tag is issued for foundation seed by seed certification agency.

iv. Certified seed

Certified seed is the progeny of foundation seed which is produced by registered seed growers or societies under supervision of qualified technical officers of certification agency in such a way that its quality is maintained according to prescribed seed standards. An azar blue colour tag is issued for certified seed by seed certification agency.

Seed Chain

Seed chain is generation system of seed multiplication for the production of a particular class of seed from specific class of seed up to stage of certified seed. The choice of a proper seed multiplication system is the key for success of a seed production programme. The selection of a seed multiplication chain depends on rate of genetic deterioration, seed multiplication ratio and total seed demand. The chain of seed multiplication models may be of three, four or five generations as mentioned below:

a) Three generation system

Breeder seed → Foundation seed → Certified seed

b) Four generation system

Breeder seed → Foundation seed I → Foundation seed II → Certified seed

c) Five generation system

Breeder seed → Foundation Seed I → Foundation Seed II → Certified Seed I → Certified Seed II

However, the production of certified seed in seed chain cannot exceed for the three generations beyond Foundation seed stage I or four generation beyond breeder seed.

Eligibility for Certification of Crop Varieties

The crop varieties which are notified under Section 5 of the Seeds Act, 1966 are only eligible for seed certification.

Certification Agency

Seed certification is conducted by the Certification Agency notified under Section 8 of the Seeds Act, 1966.

Recognized seed testing laboratories

Seed testing for physical seed quality parameters are made in recognized seed testing laboratories created under clause 4(1) of the Seeds Act, 1966.

4

Seed Sampling for Testing

Seed Sampling

The size and quantity of seed sample required for seed quality testing remains small as compared to the seed lot hence, representative samples are randomly drawn for this purpose. The process of obtaining a uniform and representative sample from a seed lot for seed quality testing is called seed sampling. The samples are prepared in accordance to ISTA rules to ensure that small size sample should represent truly and in the same proportion all constituents of seed lot.

Objective

i. To get a uniform and representative sample from a homogeneous seed lot following ISTA rules for seed physical purity analysis.

ii. To send the seed sample to recognized seed testing laboratory for seed purity test.

Time of Seed Sampling

The uniform and representative sample from a seed lot is drawn one to two months prior to the start of the sowing season.

Methods of sampling

i. Hand sampling

ii Sampling with triers or probe

i. Hand sampling

This is followed for sampling the non-free flowing seeds or chaffy and fuzzy seeds such as cotton, tomato, grass seeds etc. In this method it is very difficult to take samples from the deeper layers of bag. To overcome this, bags are emptied completely or partly and then seed samples are taken. While removing the samples from the containers, care should be taken to close the fingers tightly so that no seeds escape.

ii. Sampling with triers or probe

The samples can be drawn from bags or from seed bulk by using appropriate triers. The triers are used for taking free flowing seed samples.

Bin samplers

The samples from the seed lots stored in the bins are taken by using bin samplers.

Nobbe trier

This is suitable for sampling of the seeds stored in bag but not in bulk. It has a pointed tube long enough to reach the center of the bag with an oval slot near the pointed end. Its length is very small and it was named after the father of seed testing Fredrick Nobbe.

Sleeve type triers or stick triers

It is the most commonly used trier for sampling. These are of two types i.e. with compartments and without compartments. This trier is used for drawing seed samples from the seed lots packed in bags or in containers.

Seed Sampling

Seed lot is a physically identifiable homogeneous quantity of seed of a variety stored in bag or in bulk. A uniform and representative sample is collected randomly from seed lot in the process of seed sampling by using above mentioned methods. The drawn samples are mixed to form a composite sample. The quantity of composite seed samples shall be sufficient to provide three samples of the equal size of submitted sample. The composite sample will be divided into three equal parts, and one shall be sent for analysis to a notified Seed Testing Laboratory, the second part to the seed producer and the third part is retain as a guard sample.

The stored quantity of seed per lot depends on the seed size of the crop variety and prescribed as mentioned below:

Seed size	Maximum lot size (kg)
Larger than wheat and paddy	20,000 kg
Smaller than wheat and paddy eg. Fenugreek, Senji etc.	10,000 kg
Maize	40,000kg

Each seed lot shall be assigned a specific number in order to facilitate maintaining its identity, tracing back to its origin, handling in stores, transit etc., accounting and inventory maintenance and referring/communicating about a certain quantity of seed.

Types of Samples

i. Primary sample

ii. Composite sample

iii. Submitted sample

iv. Working sample

i. Primary sample

Each probe or handful of sample taken either in bag or in bulk is called primary sample.

ii. Composite sample

All the primary samples drawn are combined and thoroughly mixed together in suitable container to form a composite sample.

iii. Submitted sample

When the composite sample is properly reduced to the required size that to be submitted to the seed testing laboratory, it is called submitted sample. Submitted sample of requisite weight is obtained by repeated halving or by abstracting and subsequently combining small random portions.

iv. Working sample

It is the reduced sample with required weight obtained from the submitted sample after repeated mixing and dividing with which the seed quality tests are conducted in seed testing laboratory

The details are to be attached with the samples are listed below:

i. Name and address of seed sender

ii. Kind of seed and variety

iii. Class of seed

iv. Year of harvest or seed lot number in accordance to ISTA norms

v. Type of test required like

a. Germination

b. Physical purity of the seed

c. Viability or Tetrazolium

d. Moisture content in seed

e. Weed seeds

f. Inert matter

Then it can be send to the seed laboratory for seed quality analysis.

Sampling Intensity

The sampling intensity or number of required primary samples should be maintained as ISTA norms as mentioned below:

a. For seed lots in bags or container of similar capacity that is uniform in size

Up to 5 containers	Sample each container but never < 5 primary samples
6-30 containers	Sample atleast one in every 3 containers but never < than 5 primary samples
31-400 containers	Sample atleast one in every5 containers but never < than 10 primary samples
401 or more containers	Sample atleast one in every 7 containers but never < than 80 primary samples

When the seed is in small containers such as tins, cartons or packets, a 100 kg weight is taken as the basic unit and small containers are combined to form sampling units not exceeding these weight e.g. 20 containers of 5 kg each. For sampling purpose each unit is regarded as one container.

b. For seeds in bulk

Up to 500 kg	At least 5 primary samples
501 to 3000 kg	Primary samples for each 300 kg but not less than 5 primary samples
3001-20,000 kg	Primary samples for each 500kg but not less than 10 primary samples
20001 kg and above	Primary samples for each 700 kg but not less than 40 primary samples

Weight of Submitted Sample

The minimum weight for submitted samples for various tests is as follows:

a. Moisture test

100 grams for those species that have to be ground and 50 grams for all other species.

b. For verification of species and cultivar

Crop	Lab only (g)	Field plot & lab (g)
Peas, beans, maize, soybean and crop seeds of similar size	1000	2000
Barley, oats, wheat and crop seeds of similar size	500	1000
Beetroot and seeds of similar size	200	500
All other genera	100	200

c. For other tests like purity and count of other species

Crop	Size of seed lot (kg)	Size of submitted Sample (g)	Size of working sample for purity (g)	Size of working sample for count of other species (g)
Coriander	10,000	400	40	400
Fennel	10,000	180	18	180
Fenugreek	10,000	450	45	450

The samples taken may be packed in bags, sealed and marked for identification. For moisture testing the samples should be packed separately in moisture proof polythene bag and kept in the container along with submitted samples.

Types of sample used in seed testing lab

Service sample	:	Sample received from the farmers.
Certified sample	:	Sample received from certification agencies or officers.
Official sample	:	Sample received from the seed inspectors.

Mixing and Dividing of Seeds

Mechanical dividing

The reduction of sample size is carried out by the mechanical dividers suitable for all seeds except for chaffy and fuzzy seeds.

Soil divider: It is suitable for large seeds and chaffy seeds.

Random cup method

This method is suitable for seeds requiring working sample up to 10 grams provided that they are not extremely chaffy and do not bounce or roll (e.g) *Brassica* spp. Six to eight small cups are placed at random on a tray. After a preliminary mixing, the seed is poured uniformly over the tray. The seeds that fall into the cup is taken as the working sample.

Spoon method

This is suitable for samples of single small seeded species. Sufficient portions of seed are taken to estimate a working sample approximately but not less than the required size.

Hand halving method

This method is restricted to the chaffy seeds. The halved portions are arranged in rows and alternate portions are combined and retained. The process is repeated until the sample of required weight is obtained.

Dispatch of Submitted Sample

i. Each submitted sample should be sealed and marked.

ii. The label should contain all the necessary details as listed above.

iii. The seed should be packed to avoid damage during transit and in cloth bag for germination test. It should be packed separately in moisture proof containers for moisture content determination.

iv. Samples should be dispatched by the sampler to the seed testing laboratory without delay.

5

Seed Standards

1. Isolation distance

S. No.	Name of Seed Spice Crop	Isolation distance (m)	
		Foundation seed	Certified seed
1	Coriander	200	100
2	Fenugreek	50	25
3	Cumin	800	400
4	Fennel	200	100
5	Ajowain	200	100
6	Dill	200	100

2. Field inspection

S. No.	Name of Seed Spice Crop	Number of field inspection	Crop stage	Inspection points
1	Coriander Fenugreek Cumin Fennel Ajowain Dill	3	1st : Before flowering preferably within 45 days of sowing 2nd : During 50% flowering 3rd: At maturity and prior to harvesting	Determine isolation, volunteer plants, off-types To check isolation and off-types To verify the true nature of plant

3. Off type plants

S. No.	Name of Seed Spice Crop	Maximum permitted limit of off type plants (%)	
		Foundation seed	Certified seed
1	Coriander	0.10	0.50
2	Fenugreek	0.10	0.50

3	Cumin	0.05	0.10
4	Fennel	0.10	0.50
5	Ajowain	0.10	0.50
6	Dill	0.10	0.50

4. Objectionable weeds

S. No.	Name of Seed Spice Crop	Name of objectionable weed	Maximum permitted limit of objectionable weeds (%)	
			Foundation seed	Certified seed
1	Coriander	-	None	None
2	Fenugreek	Melilotus (Senji)	0.01	0.02
3	Cumin	Plantago pumila (jiri)	0.05	0.10
4	Fennel	Cuscuta spp.	None	0.05
5	Ajowain	Trachyspermum roxumburgh	0.01	0.02
6	Dill	-	None	None

5. Seed purity standards

i. Foundation seed

S. No.	Name of Seed Spice Crop	Seed purity standards (%)					
		Germination (%)	Moisture (%)	Pure seed (Minimum)	Inert matter (Maximum)	Other crop seed (Maximum)	Weed seed (Maximum)
1	Coriander	65	10	97	3	10/kg	None
2	Fenugreek	70	8	98	2	10/kg	10/kg
3	Cumin	65	10	97	3	10/kg	10/kg
4	Fennel	65	10	97	3	10/kg	10/kg
5	Ajowain	65	10	95	3	10/kg	10/kg
6	Dill	65	10	97	3	10/kg	10/kg

ii. Certified seed

S. No.	Name of Seed Spice Crop	Seed purity standards (%)					
		Germination (%)	Moisture (%)	Pure seed (Minimum)	Inert matter (Maximum)	Other crop seed (Maximum)	Weed seed (Maximum)
1	Coriander	65	10	97	3	20/kg	None
2	Fenugreek	70	8	98	2	20/kg	20/kg
3	Cumin	65	10	95	3	20/kg	20/kg
4	Fennel	65	10	97	3	20/kg	20/kg
5	Ajowain	65	8	95	5	20/kg	20/kg
6	Dill	65	10	97	3	20/kg	20/kg

NB: I. The moisture % for vapour proof container should be 2% less than the prescribed limit.
II. Insect damaged seed should not be more than 5% in each sample.

6

Physical Purity Test

Physical purity test determines the proportion of pure seed, other crop seed, weed seed, inert matter, moisture and germination components in the seed lot by weight in percentage in accordance to prescribed Seed Standards. It thus, helps in improving the plant stand, raising a pure crop by eliminating other crop seeds, weed seeds and disease free-seeds. Seed must not be sold until the percentage of pure seed, other crop seed, weed seed, and inert matter are in accordance to seed physical purity standards. Physical purity test is required for seed certification or seed law enforcement and it is a pre-requisite for germination test.

Objective

i. To determine the percentage composition of pure seed and other prohibited matters in the sample being tested for seed certification and seed law enforcement.

ii. To know the status of seed quality suitable for sowing.

iii. To assess the labeling specifications of seed quality parameters.

Importance

- It determines the planting value.
- It avoids the hazards of crop production by furnishing the information on different quality standards like purity, moisture, germination, vigour and health.
- It determines the genuineness of crop varieties and authenticity of the certified seed lot.
- It helps in quality control of seed lot which is offered for sale.

Pure Seed

Pure seed means seed of spice crop free from inert matter, weed seeds and all other seeds distinguishable from a kind or variety being considered or under limit of prescribed Indian Minimum Seed Standards. The hard, undersized, immature, shriveled diseased or germinated seeds are also pure seeds.

Other Crop Seed

Other crop seed means the seed crops or plant species other than kind being examined or units of any other than of pure seed grown as crops.

Weed Seeds

Weed seed means the seeds of those species normally recognized as weeds or specified under Seed Act as a noxious weed.

Inert Matter

Inert matter means the seed like structures and all other matter not defined as pure seed excluding other crop seed and weed seeds. It includes seed like structures, stem pieces, leaves, sand particles, stone particles, empty glumes, lemmas, paleas, chaff, awns, stalks longer than florets and spikelets.

Equipment used for Purity Analysis

i. Seed blower

ii. Purity work board

iii. Counting board

iv. Vacuum counter

v. Digital moisture meters

vi. Seed germinator

vii. Room germinator

viii. Impression board

ix. Electronic weighing balance

i. Seed blower

Seed blowers are the two plastic columns, one for larger seeds and the other for smaller seeds provided with a semi-circular outlet where the terminal velocity of wind can be adjusted. A time clock is also provided for the automatic running of the bowler. The inert matter is separated by stratification using the terminal velocity of air.

ii. Purity work board

It is used for separation of different components of seed lot. At the center of the board, there is an illumination by which the emptiness of the seed is easily identified.

iii. Counting board

It is used for accurate counting and spacing of seeds. This consists of 2 plates. The basal one is stationary and top one is movable. Both top and basal plates are having uniform number of holes *viz.*, 50/100,when the plates are in different position. After taking the sample, the top plate is pulled in such a way that the holes are in one line so that the fixed numbers of seeds fall on the substratum.

iv. Vacuum counter

It consists of head, pipe and wall. The plates bear 50 or100 holes which can be fitted to the head. When vacuum is created the plate absorbs seeds and once the vacuum is released the seeds fall on the substrate.

v. Digital moisture meters

It estimates seed moisture quickly. It works on principle that moisture content is directly proportional to the electrical conductivity of the seed. Because moist grains are good conductors while dry grains are less conductors of electricity. It has a compression unit to compress the sample to predetermined thickness. The thickness setting is very easily read on a vertical and circular scale. The seed material on test is taken in a test cup and is compressed. Then press the push type switch till the reading comes in the display. The computer versions of digital moisture meter automatically compensate for temperature corrections.

vi. Seed germinator

It is closed chamber where temperature and relative humidity are controlled. Thus, maintenance of required temperature can be made easily. Room germinator works with same principles of seed germinator. This is modified chamber of larger one and the worker can enter into it and evaluate the seedlings. Provisions are made to maintain the temperature and relative humidity. This is used widely in practice.

vii. Impression board

This is made of plastic / wood with 50 or 100 holes per pins. Here the knobs are arranged in equal length and space. By giving impression on the sand it makes uniform depth and spacing for seed.

viii. Electronic weighing balance

It is an electronic machine used to measure the weight of seeds in the laboratory. These balances come in precision designs and their operating characteristic allows the quick and accurate measurements. Further, these balances can also be tubes to transfer data to computer.

Analysis of Seed in Laboratory

Seed testing is important for the producer, sellers and seed users. Seed testing is highly specialized and technical job. Seed purity analysis is carried out by testing the cleanness, testing of geneuiness of the variety, moisture testing and testing of viability and germination.

The seed samples received in the laboratory (submitted sample) are reduced to obtain working samples for carrying out the various components of seed purity tests. The reduction of sample size is carried out by a number of methods as mentioned below:

i. Mechanical divider

ii. Modified halving method

iii. Hand halving method

iv. Random cup method

v. Spoon method

Working Sample

The purity analysis is carried out on the working sample of prescribed weight drawn from submitted samples. The analysis may be made on one working sample of the prescribed weight, or on two sub-samples of at least half this weight, each drawn independently.

Weighing of working sample

The number of decimal places to which the working sample and the components of the working sample should be weighed are mentioned below:

Weight of the working sample in gram	Number of decimal places required	Example
<1	4	0.7534
1-9.999	3	7.534
10-99.99	2	75.34
100-999.9	1	753.4
1000 or more	0	7534

Physical Separation

The working sample after weighing is separated into its components *viz*., pure seed, other crop seed, weed seed and inert matter.

Method of Purity Separation

First of all clean the seed sample by sieving / blowing operations. Then place the seed sample on the purity work board and separate into its components like other crop seeds, weed seeds and inert matter. After separation identify each kind of weed seeds, other crop seeds as to genus and species. The names and number of each are recorded. The type of inert matter present should also be noted.

Determination of Components in Percentage

All the four components must be weighed to the required number of decimal places. The percentage of the components is determined by using the following formula:

$$\%\text{ of component} = \frac{\text{Weight of individual component}}{\text{Total weight of all components}} \times 100$$

Another analysis is made if there is a gain or loss between the weight of the original samples and the sum of all the four components, which is in excess of one percent analysis results are nearer to borderline of the seed standards, one more test is carried out and the average result is reported. However, if a duplicate analysis is made of two half samples, or whole samples, the difference between the two must not exceed the permissible tolerance.

Determination of Inseparable other Crop Seeds and Objectionable Weed Seeds

Whole submitted sample is used for determination of inseparable other crop seeds and objectionable weed seeds. The number of inseparable other crop seeds per kg or number of objectionable weed seeds per kg is counted and reported even if the working sample is less than one kilogram.

Determination of other distinguishable varieties (ODV)

The size of sample used for the determination of other distinguishable varieties remains ten times than the size of working sample is used. ODV is determined based on the morphological characters of the seeds and the authentic samples remain available for comparison. The number of ODV is determined and reported as number per kilogram of seeds.

Calculation of results

The percentage by weight of each of the component is calculated at one decimal place.

Reporting results

The results of each component are reported in one decimal place. The total of all components must be 100. Components of < 0.05% is reported as Trace. If the result for a components is nil, this must be shown as '-0.0-' in the appropriate space. The procedure for determination of germination percentage and moisture content in seeds are described in separate chapter.

7

Germination Test

Seed Germination Test

Germination is the process of emergence and development of seed embryo into a plant under favourable environmental conditions of water, soil, sunlight and temperature. The best example of germination is the sprouting of a seedling from a seed of an angiosperm plant. It reflects the ability of seed to produce a normal plant under favourable environmental conditions. The germination rate of a particular seed lot is a key indicator of its performance in the field.

Objective

i. To know the emergence and development potential of seed lot into a plant under favourable environmental conditions of water, substrate, sunlight and temperature.

ii. To determine the germination rate of a particular seed lot to know its performance in the field.

Principles

A germination test is conducted with the pure seed fraction. A minimum of 400 seeds are required, which can be replicated in four, eight or sixteen replicates of 100, or 50 or 25 seeds each. The number of replications depend on the size of the seeds and size of containers of substrate. The favourable conditions of moisture, temperature, suitable substratum and light are necessary to conduct the test. No pre-treatment to the seed is given except for those recommended by ISTA.

Types of Germination

Two types of seed germination occur which are

I. Epigeal germination

II. Hypogeal germination

Both type of germination are not related to seed structure. The seeds of bean and pea are mostly similar in structure but their germination patterns are quite different. The germination in beans is epigeal while, this is hypogeal in peas.

Epigeal germination

During epigeal germination, the cotyledons are raised above the ground. During root establishment, the hypocotyl begins to elongate in an arch which breaks thro' the soil, pulling the cotyledon and enclosed plumule (epicotyl) thro' the ground and projecting them in the air.

Hypogeal germination

During hypogeal germination, the cotyledons remain beneath the soil while the plumule pushes upward and emerges above the ground. Here the epicotyl (plumule) elongates.

Requirement for Germination Test

I. Substratum

II. Temperature

III. Light

IV. Germination apparatus

V. Seed counting boards

I. Substratum

The substratum provides a surface or medium for the germination of seeds and seedlings grow. It also serves as moisture reservoir. Sand, paper and soil are the commonly used substrata.

i. Sand

Sand particles, used as substratum should not be too large or too small. The sand particles should pass through 0.80 mm sieve and retained by 0.05 mm sieve. Sand should not have any toxic material or any pathogen. The sand should be sterilized in an autoclave, if any pathogen is found.

Germination tray

Germination trays are used to carryout the germination test, when sand is used as substratum. The normal size of the tray is 22.5 x 22.5 x 4 cm. The tray may be either of zinc or stainless steel.

Method of Seed Placement in Tray

Seeds are planted in a uniform layer of moist sand and then covered to a depth of 1 cm to 2 cm with sand. Seeds are pressed into the surface of the sand. The equal spacing from all sides should be given to facilitate normal seedling growth, to avoid entangling of seed and spread of disease. Spacing should be 1-5 times of the width or diameter of the seed.

The amount of water to be added to the sand will depend on size of the seed. Generally, the sand can be moistened to 50% of its water holding capacity in small seeded crops like seed spices.

ii. Paper

Most widely used substrates for seed germination are filter paper, blotter or towel/ kraft paper, which possess ability for capillary movement of water at vertical direction at the rate of 30 mm rise / min. The paper should be free from toxic substances, fungi and bacteria and possess ability to holds sufficient moisture during the germination period. The texture should be such that the roots of germinating seedlings will grow on and not into the paper.

Methods of seed placement on the paper

The seed for germination is placed either on moist paper or between two layers of the paper. Accordingly, the methods are called as top of paper (TP) or between paper (BP).

Top of paper (TP)

Seeds are placed on one or more layers of moist filter paper or blotter paper in petri dishes. The petri dishes are covered with lid and placed inside the germination cabinet. This is suitable for those seeds, which require light.

Between paper (BP)

The seeds are placed between two layers of paper for germination. After this, the paper is rolled in towel for its placement in either germinator or glass plates. Accordingly, this method is termed as roll towel method.

Roll towel method

The seeds are placed between two layers of paper and rolled in towels. The rolled towels are placed in the germinator in an upright position.

Inclined plate method

Germination on glass plate with germination paper and kept at an angle of 45°.

iii. Soil

The soil should be non-caking, free from any large particles. It must be free from weed seeds, bacteria, fungi, nematode or toxic substances. Soil is not recommended for reuse.

II. Temperature

The optimum temperature for the germination of seed spices is 20-30°C.

III. Light

Light should be provided to seeds requiring light for germination. In seed spices, fennel and coriander require light for germination. Coriander requires seven days for breaking the dormancy while, fennel and fenugreek needs light for breaking the dormancy.

IV. Germination apparatus

Germination cabinet/Germinator

These are closed chamber where temperature and relative humidity can be controlled. Temperature, relative humidity and light can be maintained in these chambers which are required.

Room germinator or walk in germinator

The room germinator or walk in germinator is also works with same principles of germinator. This is modified chamber of larger size where worker can enter into it and evaluate the seedlings. Provisions are made to maintain the required temperature, relative humidity and light. This is widely used.

V. Seed counting boards

Counting board

This is used for accurate counting and spacing of seeds. This consists of 2 plates. The basal one is stationary and top one is movable. Both top and basal plates are having uniform number of holes *viz*., 50/100, when the plates are in different position. After taking the sample, the top plate is pulled in such a way that the holes are in one line so that the fixed numbers of seeds fall on the substratum.

Vacuum counter

It is consists of a head, pipe and wall. There are plates of 50 or100 holes which can be fitted to the head. When vacuum is created the plate absorbs seeds and once the vacuum is released the seeds fall on the substrate.

Impression board

This is made of plastic or wood with 50 or 100 holes per pins. Here the knobs are arranged in equal length and space. By giving impression on the sand it makes uniform depth and spacing for seed.

The germination requirement for seed spices is summarized below:

Germination requirements for different crops

Crop	Substratum	Temp. (0 C)	Light	First count days	Final Count days	Recommen-dation for breaking dormancy
Coriander (*Coriandum sativum* L)	TP,BP	20-30, 20	L	5	10	7
Fennel (*Foeniculam vulgare* Mill)	TP,BP,TS	20-30	L	5	14	Light
Fenugreek (*Trigonella foenum graecum* L)	TP,BP	20-30, 20	-	4	14	Light
Cumin (*Cuminum cyminum*)	TP,BP	20-30	-	5	7	-

TP- Top of paper, BP- Between paper and TS- Top of Sand

Seedling evaluation

The following types of seeds and seedlings are evaluated during the germination test.

- Normal seedlings
- Abnormal seedlings
- Hard seeds
- Fresh seeds
- Dead seeds

ISTA has classified the seedlings into different categories based on the development of essential structures.

Categories of Seedlings

Normal seedlings

Seedlings that possess essential structure that is indicative of their capacity to develop into a normal plant under favourable conditions of soil, water, temperature and light.

Characters of normal seedlings

- A well-developed root system with primary roots. Some species of graminae produced seminal root or secondary root.
- The shoot axis in seedlings of epigeal germination consists of elongated hypocotyl.
- Developed epicotyl in seedlings of hypogeal germination.
- Presence of one cotyledon in monocot and two in dicots.
- A well-developed coleoptile in graminae containing a green leaf.
- A well-developed plumule in dicotyledons.
- Seedlings with primary root with limited damage but well developed secondary roots in leguminosae, graminae, cucurbitaceae and malvaceae are taken as normal seedlings.

Abnormal Seedlings

Seedlings showing some form of growth but do not have the capacity to develop into a normal plant after sowing in favourable conditions of soil, water, temperature and light.

Type of abnormal seedlings

Damaged seedlings

Seedlings exhibiting loss or badly damaged portion in any one of the essential structures and its balanced growth is not expected. Seedlings with no cotyledons, with splits, cracks and lesions on essential structures and without primary root are also considered as damaged seedlings.

Deformed seedlings

Seedlings showing weak or unbalanced development of essential structures like spirally twisted or stunted plumule or hypocotyl or epicotyl, swollen shoot, stunted roots etc., can be taken as deformed seedlings.

Decayed seedlings

Seedlings showing diseases or decayed symptoms on any one of the essential structures as a result of primary infection from the seed that prevents the development of the seedlings.

Hard seeds

Seeds do not absorbing moisture till the end of the germination test period and remain hard like the seeds of leguminosae and malvaceae are hard seeds.

Fresh seed

Seeds which are neither hard nor have germinated but remain firm and apparently viable at the end of the test period can be grouped into fresh seeds.

Dead seeds

Seeds at the end of the test period are neither hard nor fresh or have produced any part of the seedlings. Often dead seeds collapse and a milky paste come out, when pressed at the end of the test.

Re-testing

The results of unsatisfactory germination test are not reported. It needs verification of the result by second germination test that should be made by the same or alternative methods under the following circumstances.

1. Replicates performance is out of tolerance
2. Results being inaccurate due to wrong evaluation of seedlings or counting or errors in test conditions
3. Dormancy persistence or phytotoxicity or spread of fungi or bacteria. The average of the two tests shall be reported.

Use of Tolerances

The tolerance Table is used to decide the result of germination test. The result can be relied only when difference between highest and lowest replicates is within accepted tolerances.

Reporting Results

The average over seed replicates is the result of the germination test. This is reported in terms of normal seedlings percentage. The percentage is computed nearest whole number. The percentage of abnormal seedlings, hard, fresh and dead seeds is determined in the same way. The result is reported on the analysis certificate under appropriate space.

8

Seed Viability Test

Seed Viability Test

Germination test for a seed lot is a pre-requisite to know its potential of emergence under field condition. However, this test requires few days to weeks and a month in some cases. Tetrazolium test gives quick information on seed viability or germinability. It makes possible to distinguish red coloured living parts of seeds from the colourless dead ones. Staining of seeds determines viability. Completely stained seeds are viable while, partially and comparatively unstained seeds are non-viable. This test is not valid as compared to seed germination test.

The accuracy of the tetrazolium test depends on

i. A sound knowledge of the seed and seedling structures

ii. Mechanisms of the reactions leading to the staining of the seed

iii. Ability to interpret the staining pattern by critically examining the seed topography and visible seed qualities

iv. Experience gained through keen interest and natural skill.

Equipment required for conducting Tetrazolium test

1. Staining dished like watch glasses or petri dishes,
2. Beakers of 100 to 250 ml capacity
3. Single edge razor blade
4. Needles
5. Forceps
6. Magnifying devices
7. Medicine dropper
8. Dispensing bottle

9. Germinating substrate
10. Oven or incubator

Method of Tetrazolium testing

A representative sample of 50 or 100 seeds is usually sufficient but, 200 seeds in replicates of 100 seeds are recommended.

Preparation of solutions

The concentrations of tetrazolium solution used for viability test ranged from 0.1 to 100% for comparable results. Generally, 1.0% solution is used for seeds that are not bisected through the embryo while, 0.1% solution is used for seeds in which the embryo is bisected. To prepare 1.0% solution, one gram of tetrazolium salt is dissolved in the distilled or tap water to make 100ml. The pH of the solution should between 6 to 8 for best staining. Solutions of pH4 or lower will not stain even viable embryos and solution of pH more than 8 will result in too intense staining.

If the pH of the water is not in the natural range, the tetrazolium salt should be dissolved in a phosphate buffer solution. The buffer solution is prepared as follows:

Solution 1: Dissolve 9.078 g of potassium dihydrogen phosphate (KH_2PO_4) in 1000 ml of water.

Solution 2: Dissolve 11.876 g of disodium hydrogen phosphate ($Na2HPO_4.2H_2O$) in 1000 ml of water.

Take 400 ml of solution 1 and 600 ml of solution 2 and mix them together. In a litre of buffer solution prepared as above dissolve 10 grams of tetrazolium salt. This gives 1% tetrazolium solution of pH 7.0. This may be further diluted to give lower concentrations. The solution should be stored in brown bottle to prevent deterioration from light

Methods of tetrazolium testing

The seeds are first prepared for staining, then stained and evaluated in light.

Method 1

Soak the seeds in water for 3 to 4 hours. Bisect the seeds by cutting longitudinally thus exposing the main structures of the embryo. Use half of the each seed for testing example maize, sorghum, small grains, and large seeded grasses.

Method 2

The seeds are cut laterally near the center of the seed above the embryo. Place embryo end in tetrazolium solution example small seeded grasses.

Method 3

Puncture the seeds by piercing through the seed into the endosperm near the embryo but avoid injury to the embryo example small seeded grass.

Method 4

Soak the seeds in water for 3-4 hours, then remove the seed coats and place the seeds in tetrazolium solution. In some crops like cotton a thin membrane adhering to the cotyledons is also removed in addition to the seed coat.

Method 5

Seeds of soybeans and other large-seeded legumes may swell so rapidly and irregularly when placed directly in water or in tetrazolium solution and results in bursting of the seed coats. Hence, it is preferably to condition these seeds slowly in moist paper towels overnight before staining, so that they absorb moisture slowly without any damage to the seed.

Method 6

Seed coats of these seeds are permeable to tetrazolium and the embryos usually will stain without conditioning, example small seeded legumes.

Staining

The prepared seeds are placed in suitable watch glass, beakers, petri dishes bearing sufficient tetrazolium solution (1.0%) to cover the seeds and to allow its absorption. Place the containers in an incubator at dark warm conditions with sufficient solution at 40°C. The staining time varies for different kinds of seeds, different methods of preparation and different temperatures (>1 hr to 8 hrs). When sample has stained sufficiently the tetrazolium solution can be discarded and the seed sample covered with water immediately. Seed samples can also be kept for 3 days at 10°C for interpretation.

Evaluation of samples

A normal stain appears cherry red.

Monocots - Non viable

1. All structures unstained
2. Shoot largely unstained.
3. Scutellar node unstained.
4. Major areas of coleoptile unstained
5. Central area of scutellum unstained
6. Insect, mechanical or other injuries causing essential structures non-functional.

Dicot - Non viable

1. Embryo completely unstained
2. More than extreme tip of radicle unstained
3. More than half of cotyledon tissue unstained
4. Deep seated necrosis at cotyledon and embryonic axis juncture or on radicle
5. Fractured radicle

Advantages of Tetrazolium test

1. Quick and fairly accurate
2. It can also determine the viability of a dormant seed lot in short time.
3. Seeds are not damaged and can be germinate

Disadvantages of Tetrazolium test

1. Distinction between normal and abnormal seedlings is difficult
2. It cannot differentiate between dormant and non-dormant seeds
3. Correct evaluation is possible only after prolonged experience
4. Harmful microorganisms for seedling emergence remain undetected.

9

Seed Moisture Test

The seed moisture content is the amount of water present in the seed. It is generally expressed in percentage of dry and wet seed weight. It is the loss in weight of the original sample when it is dried. High moisture content in seed sample favors the mechanical damage, seed longevity, incidence of disease and insect pests.

Objective

i. To know the moisture content of seeds by suitable methods

ii. To suggest the seed drying.

Methods of Moisture Estimation

i. Air oven method

ii. Moisture meters

i. Air oven method

In this method, seed moisture is removed by drying at a specified temperature for a specified duration.

ii. Moisture meters

Moisture meters estimate seed moisture quickly but the estimation is not as precise as by the air oven method.

Weight of the submitted sample

The required weight of submitted sample is100 gram for species that have to be ground while, 50 gram for all other species. The sample should be submitted in polythene bag of 700 guage.

Seed moisture determination by air oven method

Materials required

i. Grinding mill
ii. Container
iii. Oven
iv. Desiccator
v. Analytical balance
vi. Sieves

i. Grinding mill

It should be constructed of non-absorbent material. It should grind evenly and should be operated at such a speed that during grinding, it should not cause heating of the ground material. Air currents that might cause loss of moisture must be reduced to a minimum. The fineness of grinding should be adjustable.

ii. Container

Containers of glass or non-corrosive metal like stainless steel should be used.

iii. Oven

A good quality electric air oven with a thermostatic or electronic temperature control for maintaining temperature within $\pm$ 1°C is required.

iv. Desiccator

The desiccators are drying agents which are used to absorb water vapors from reagents that are hydroscopic. The most common desiccators used in laboratories are circular and made up of heavy glass. These desiccators have two compartments in which lower compartment stores a desiccator and upper compartment bear the reagent. The reagent inside the desiccator is exposed to water vapors.

v. Analytical balance

An analytical balance measures the weight of samples in grams with high accuracy. These are used when a high degree of precision is needed.

vi. Sieves

A set of wire mesh sieves with meshes of 0.5 mm, 1.0 mm and 4.0 mm is required.

Procedure of Moisture Determination

- Seed moisture determination is carried out in duplicate on two independently drawn samples.
- First weight the empty bottle / container with its cover with an accuracy of 0.01g.
- Grind the seed sample using grinder that does not cause heating or loss of water and pass through a wire sieve with meshes of 0.5 mm and not more than 10 remain on a wire sieve with a mesh of 1.0 mm.
- Mixed thoroughly the submitted sample and transfer small portions (4 or 5 gram) directly to weighing container.
- Measure the weight of sample. After weighing remove the cover of weighing bottle or container.
- Place the weighing bottle or container in an oven maintaining the desired temperature for recommended period of 1 to 4 hours at 130 – 133 ^{o}C.
- At the end of seed drying period, weighing bottle or container be closed with its cover and transfer the same to the desiccator having silica gel to cool down for 40 to 45 minutes.
- Weight again the weighing bottle or container.

Calculation of Moisture Content

The moisture content of the sample in percentage by weight is calculated to one decimal place by the following formula.

$$\text{Seed moisture content (M)} = \frac{M_2 - M_3}{M_2 - M_1} \times 100$$

Where,

M_1 =Weight of the empty container with its cover

M_2= Weight of the container with its cover and seeds before drying.

M_3= Weight of the container with its cover and seeds after drying.

The duplicate result of the determination may not differ by more than 0.2% otherwise the analysis should be repeated.

If the material is pre-dried, the moisture content is calculated from the results obtained in the pre-dried and dried stages using the following formula.

$$\text{Seed moisture contents} = \frac{(S_1 + S_2) - (S_1 \times S_2)}{100}$$

Where,

M = moisture content

S1= moisture percentage lost in pre drying stage

S2= moisture percentage lost in drying stage.

Moisture meters: Universal Osaw digital moisture meters

The principle involved in these moisture meters is that wet grains are good conductors while dry grains are less conductors of electricity. So the moisture content is directly proportional to the electrical conductivity of the seed. It consists of a compression unit to compress the sample to pre-determined thickness. The thickness setting is very easily read on a vertical and circular scale. The seed material on test is taken in a test cup and is compressed. Then press the push type switch till the reading comes in the display. Here no temperature reading and correlated dial are required. The computer versions of digital moisture meter automatically compensate for temperature corrections.

Reporting of Results

The seed moisture content is reported to the nearest of 0.1 % on ISTA certificate. If seed moisture content is determined using moisture meter the brand name and type of equipment be mentioned on analytical certificate, under the column “other determinations” reporting of range for which the moisture meter is calibrated.

10

Seed Storage

Seed storage is the preservation of viable seeds from the time of collection to the required for sowing. It is important to get adequate plant stands in addition to healthy and vigourous plants.

Objectives

i. To maintain initial seed quality during storage period by providing suitable conditions.

ii. To maintain the seed germination and emergence rates.

iii. To prevent the seed deterioration during storage period.

iv. To maintain the physical and physiological seed qualities.

Factors Affecting Seed Longevity in Storage

A. Genetic factors

- Seed morphology
- Seed physiology
- Seed maturity
- Seed microflora

B. Environmental factors

- Seed moisture during storage
- Temperature during storage
- Relative humidity in atmosphere

Storage Facilities

- Traditional storage systems
- Seed beans

- Seed containers
- Moisture controlled containers
- Seed godowns
- Cold storage
- Seed banks

Sanitation in Seed Godown

- Storage environment should be free from microbes.
- Seed godown should be kept cool and dry.
- Fumigation may be done whenever needed.
- Wooden pallets should be used for arranging the bags in cris-cross manner. It helps in effective ventilation from all sides of the bags.
- Seed bags should be stacked upto 5-8 tires depending upon density of seeds.
- Restacking once in 3 months or less is important for prolonging seed viability.
- Spraying of insecticides should be made in godown before storage of the seed.
- Cloth bags and containers should be fumigated before storage.
- Size of the stack should be 30 x 20 feet to facilitate fumigation under gas proof or polythene covers.
- Chemicals such as insecticides, fertilizers etc should not be stored along with seeds.

Maintenance of Viability in Storage

- Store well mature seeds.
- Store normal coloured seeds.
- Seeds should be free from mechanical injury.
- Seeds should be free from storage microorganisms.
- Seeds should be treated with fungicides before storage.
- Seeds should not have met with adverse conditions during maturation.

- Storage environment or godown should be dry and cool.
- Seeds should be dried to optimum moisture content before storage.
- Required relative humidity and temperature should be maintained during storage.
- Storage godown should be fumigated to control storage insects, periodically.
- Suitable packaging materials should be used for packing.

11

Formats for Seed Analysis Report

Mini seed testing lab analytical register

S. No.	Name and address of the applicant	Crop and Vari-ety	Class of the seed	Details of the seed source	Date of expiry	Quan-tity of the lot	Date of receipt	Date of sam-pling
1	2	3	4	5	6	7	8	9

Date of test	Analytical result				Date of dis-patch of result	File No.	Seed rate recom-menda-tion	Sig-nature of the seed analyst
	Normal seedlings	Abnormal seedlings	Hard seed	Dead seed				
10	11	12	13	14	15	16	17	18

Seed analytical register

S. No.	Name and address of the producer	Kind / variety	Class	S.C. No	Lot No	Quantity of the lot	Date of test
1	2	3	4	5	6	7	8

Analytical result				Germination (%)	Weed seed (%)	ODV (%)	Inert matter (%)	Signature of the seed analyst	Remarks
Purity (%)	Other crop seed (%)	Hard seed	Moisture (%)						
9	10	11	12	14	15	16	17	18	19

ODV = Other distinguished variety

Seed samples register

Name of the SCO	Date of drawing sample by SCO	Date of receipt of Sample by ADSCN	ADSC Code No	Date of receipt of analytical report	Kind of sample [Fresh/ Resample/ Validation]	STL Code No	Date of sending of result to the producer	Signature of ADSCN
1	2	3	4	5	6	7	8	9